7th International Conference on Upstream Engineering and Flow Assurance 2019

Topical Conference at the 2019 AIChE Spring Meeting and 15th Global Congress on Process Safety

New Orleans, Louisiana, USA
31 March - 4 April 2019

ISBN: 978-1-5108-8678-0

Printed from e-media with permission by:

Curran Associates, Inc.
57 Morehouse Lane
Red Hook, NY 12571

Some format issues inherent in the e-media version may also appear in this print version.

Printed by Curran Associates, Inc. (2019)

For permission requests, please contact AIChE
at the address below.

AIChE
120 Wall Street, FL 23
New York, NY 10005-4020

Phone: (800) 242-4363
Fax: (203) 775-5177

www.aiche.org

Additional copies of this publication are available from:

Curran Associates, Inc.
57 Morehouse Lane
Red Hook, NY 12571 USA
Phone: 845-758-0400
Fax: 845-758-2633
Email: curran@proceedings.com
Web: www.proceedings.com

TABLE OF CONTENTS

550150 Investigating the Effect of Asphaltenes on Wax Deposition Using a Taylor-Couette Cold Finger Device

Wax deposition brings severe risks to the safe operation of waxy crude pipelines. In previous wax deposition studies, the role of heavy components in the crude such as asphaltenes was almost neglected. The impact of asphaltenes on wax deposition was systematically investigated utilizing a Taylor-Couette cold finger equipped with a rotary outer cylinder and a coaxial stationary inner cylinder. This apparatus could generate precisely controlled flow fields compared with traditional cold fingers with stirring. The experiments were carried out by using model oils with asphaltene concentrations ranging from 0wt% to 0.4wt%. The mass, thermal characteristics, composition and rheological properties of the deposits were obtained through weighting, DSC, HTGC and rheometer, respectively. The content, molecular weight and dipole moment of asphaltenes separated from deposits with n-heptanes were then measured or calculated. The results indicate the following three points. (a) the addition of asphaltenes significantly decreases the deposit mass but increases the wax content (see Figu re 1 a) and critical carbon number (CCN) of the deposit. In addition, asphaltenes suppress the increment of the shorter n-paraffin fractions in the deposit and inversely accelerate the increment of longer n-paraffin fractions. Conclusively, the presence of asphaltenes at experimental concentrations facilitates the aging process of the deposit. (b) the asphaltene content in the deposits constantly increases with deposition time elapsing. As shown in Fi gure 1b , the asphaltenes are enriched approximately 3-flod in the 1-hour deposit, 5-flod in the 12-hour deposit and over 6-flod in the 24-hour deposit compared to the oil. Moreover, both the number average molecular weight (*Mn*) and the weight average molecular weight (*Mw*) of asphaltenes separated from the deposits increase while the polydispersity index (PDI) decreases in contrast with the oil. Apparent dipole moment calculations indicate that asphaltenes extracted from the deposits are more polar than those from bulk solutions. This implies that the asphaltene components with high polarity or molecular weight are prior to entering into the deposit. Based on (a) and (b), we generalize the conception of wax deposit aging via incorporating in the enrichment of asphaltenes beyond the diffusion of paraffin molecules greater or less than CCN. (c) adding high concentration asphaltenes (more than 0.2wt% for our experiments) would contribute to the shear stripping of deposits because asphaltenes could remarkably weaken the structural strength of deposits. Taking model oil with 0.2wt% asphaltenes for example, the properties of the deposits on the cold finger prior to and subsequent to shear stripping are contrasted. The results demonstrate that deposit mass, gelation temperature and yield stress of deposits prior to stripping significantly decrease from a macro perspective, and individual carbon numbers and asphaltene content also decrease from a micro perspective compared with deposit subsequent to stripping. This study provides a fundamental understanding of the role of asphaltenes on wax deposition and a basis for modeling wax deposition of waxy crude pipelines.

This work is financially supported by the National Natural Science Foundation of China (No.51534007 and No.51134006).

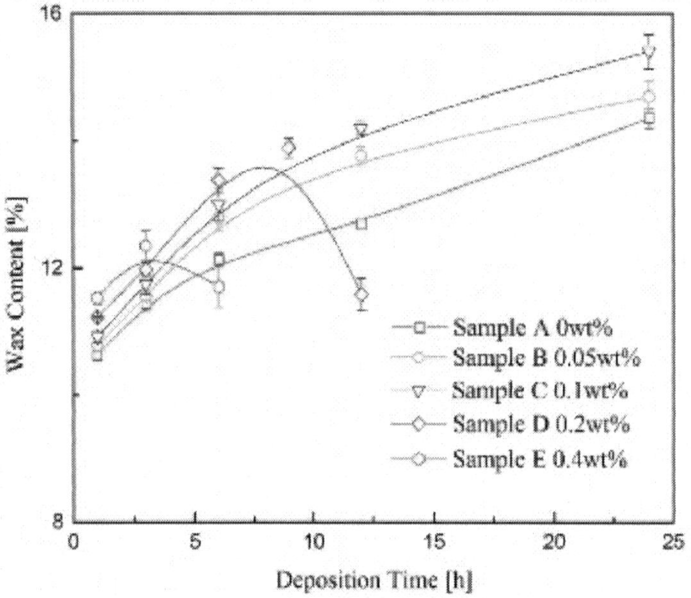

(a)

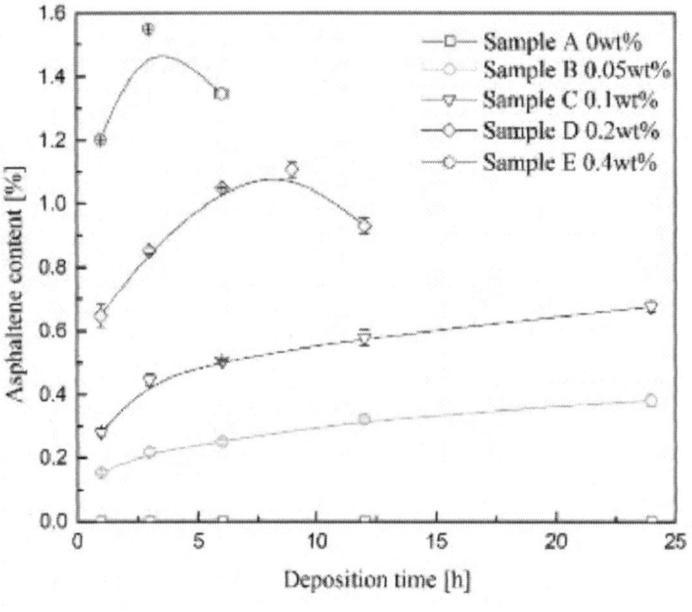

b

Figure 1. Deposit wax content and asphaltene content variation with time for model oils with different asphaltene concentrations (a) wax content (b) asphaltene content

Magnetic Treatment of Asphaltenic Oils (an MD simulation study)

Mohammed H. Khalaf [a,b,*], *G.Ali Mansoori* [c]

[a] Department of Chemical Engineering, University of Illinois at Chicago, Chicago, IL 60607-7052, USA
[b] Department of Refining, Baiji Oil Training Institute, Baiji, Iraq
[c] Departments of Bio and Chemical Engineering, University of Illinois at Chicago, Chicago, IL 60607-7052, USA

* Corresponding author.
E-mails: mkhala5@uic.edu (M. Khalaf).

Abstract

Magnetic treatment is being used in oil industry to improve crude oil rheological properties and reduce the associated heavy organics problems. The scientific basis and general conclusion of the effect of magnetic field on asphaltene are not clear. Molecular dynamics (MD) simulations are used to study the effect of magnetic fields on the onset of asphaltene aggregation. Four different model asphaltenes (A1-A4) were used. Radial distribution functions, hydrogen bond interactions, and the diffusivities were analyzed and investigated. Magnetic field intensity of 1 T could reduce the onset of aggregation of A1 and A2 by 25% and 20% respectively. In the case of A3 and A4, magnetic fields did not show high changes in the onset of asphaltene aggregation. This may be due to the flexibility of the molecules where A3 is archipelago asphaltene, and A4 has three long flexible sulfur-containing chains. Magnetic fields influenced the average number of hydrogen bonds and diffusivities according to the intensity of the field and asphaltene architecture. Asphaltenes with large single aromatic core changed more than others and the archipelago structure was less affected.

1. Introduction

Increasing worldwide demand for energy and depletion of the conventional oil sources has created a need for trapping unconventional oil resources[1]. However, using these resources is associated with several challenges such as aggregation and deposition of heavy organics (e.g., asphaltenes) and low mobility arising from their high viscosities[2–5]. Recently, magnetic fields became one of the proposed approaches to overcome these challenges[6–8]. The detailed information about the role of magnetic field in enhancing or preventing the association of heavy organic materials is still lacking.

There are several experimental and theoretical studies about the magnetic treatment of crude oils. Some of the studies claimed that polar species (e.g., asphaltenes) were highly affected by magnetic fields[9] while others focused on paraffines[10]. In paraffinic oils, some of the results found that magnetic field reduced wax crystallization. However, others reported that a magnetic field could enhance wax crystallization.

Most of the studies were about wax and there is little scientific information about the effect of magnetic fields on asphaltenes. Due to the presence of different heteroatoms in asphaltene and the asymmetrical charge distribution, we believe that asphaltenes are highly affected by magnetic fields in comparison with paraffines. It is believed that magnetic treatments change polar species intermolecular interactions which lead to form new associated structures[11]. Therefore, asphaltene molecules may aggregate or disaggregate because of magnetic field. Accordingly, we were motivated to undertake the present study. Molecular dynamics (MD) simulation technique is used to investigate the onset of asphaltene aggregation and its behavior due to magnetic field in a well-defined media.

2. Simulation methodology

Asphaltenes are the heavy and polar part of crude oils. They are soluble in aromatics and insoluble in paraffins. Asphaltenes are found in different structures, depending on the location and reservoir properties. Due to the differences in asphaltene structures, they have different properties such as aromaticity and polarity. Studying asphaltene behavior during magnetic treatments needs to consider different structures. Thus, four different asphaltenes named A1, A2, A3, and A4 were chosen and reported in Figure 1. A 7 wt% asphaltene concentration (24 asphaltene molecules) was used to represent the average asphaltene concentration found in many crude oils. In each simulation, asphaltene molecules were randomly distributed in the simulation box and placed in an equimolar mixture of n-heptane and o-xylene at standard conditions.

OPLS-AA force field was used for utilizing all the molecules in the system (asphaltenes, n-heptane, and o-xylene). OPLS-AA was shown to work well for reproducing experimental data of organic liquids. Therefore, it was used successfully for asphaltene aggregation simulations[12–17]. Topology parameters of the system and input files were generated using DL_FIELD[18]. All MD simulations were conducted using DL_POLY_4 simulation package[19]. The simulation methodology of magnetic field implementation and the simulation details can be found in the previous studies[8,20]. Four different magnetic fields (0.0, 0.5, 1.0, and 1.5), measured in Tesla (T), were applied to the system along the z-direction for 20 ns. All the collected data was analyzed using DL_ANALYSER[21]. The molecular configurations and trajectories were visualized using VMD package[22].

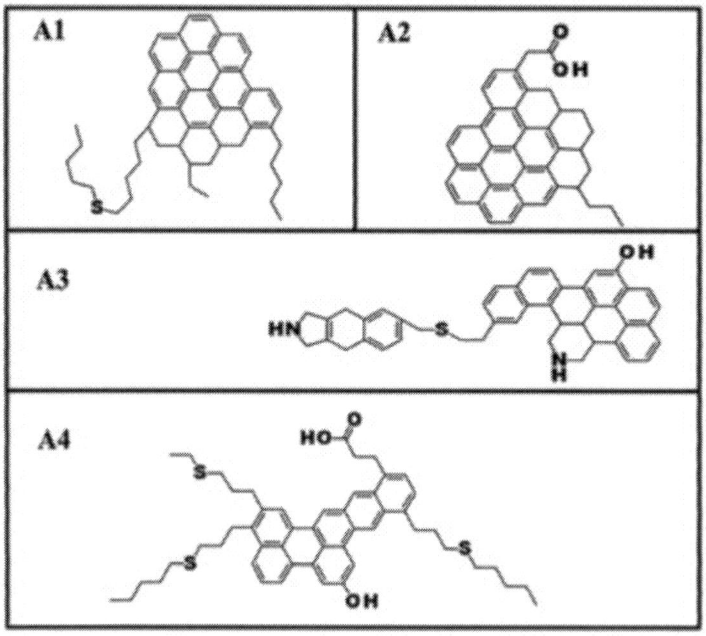

Figure 1. Model asphaltenes used in this study.

3. Results and discussions

Twenty-four molecules of each of the three kinds of asphaltenes (A1, A2, A3, and A4) reported in Figure 1 were used in each simulation. Radial distribution functions (RDFs), and hydrogen bond (HB) interactions, transport properties, and trajectories were investigated to understand the effect of magnetic fields on various asphaltene's behavior.

3.1 Radial distribution functions

(a) A1 asphaltene

In the magnetic-free environment, the RDF curve shows multiple peaks at different distances. The first clear peak is at 6 Å, and a second peak around 8.5 Å (Figure 2-A1). Trajectory inspections showed that the first peak was due to parallel-displaced or T-shaped stackings. Moreover, the second peak is due to the nearest neighbor molecules that are loosely attached to the stacked molecules. The stacking distances agreed with the reported distances from our previous works [8,12].

When the magnetic field was applied, at 0.5 T, a high RDF peak at 5 Å became visible. Inspection of molecular stacking showed that this peak was due to face-to-face stacking. In addition, this peak was more clearly defined when compared with that of the peak without the magnetic field application. This clearly indicates the enhancement of the molecular aggregation due to the magnetic field effect. Increasing the magnetic field to 1.0 T showed interesting results. It showed a clear reduction in both peaks in comparison with the magnetic-free case. In fact, by comparing the intensities of the first peaks between 0.0 T and 1.0 T, the 1.0 T field strength reduced the onset of asphaltene aggregation by about 25%. The disruption of molecular aggregates become even more evident at a higher magnetic field strength of 1.5 T.

(b) A2 asphaltene

In the absence of magnetic field, the RDF curve shows multiple peaks at different distances where they represent multiple stackings (Figure 2-A2). The first clear peak is at 4.7 Å, and a second higher and border peak is at 8.2 Å. Tracking the molecular configurations showed that the first peak was due to face-to-face stackings and the second peak was due to parallel-displaced or T-shaped stackings. It is clear that A2 produced large aggregates in the absence of magnetic field due to π-π interactions. One of the aggregates was composed of 11 asphaltene molecules in different stackings which represent asphaltene cluster as shown in Figure 5.

When the magnetic field was introduced into the system, at 0.5 T, a high RDF peak at 8.2 Å became more visible in comparison with mthe agnetic-free system. This could be explained by aggregation enhancement due to magnetic interactions. Surprisingly, applying a magnetic field of 1.0 T showed a clear reduction in the second peak in comparison with mthe agnetic-free system. Comparing RDF peaks of magnetic-free system and 1.0 T field strength showed that 1.0 T could eliminate the aggregation by ~20%. Increasing magnetic field strength to 1.5 T showed negative results in reducing asphaltene aggregation. It could be seen that higher magnetic field increased the aggregation by 30% and changed the dominated stacking to be face-to-face stacking.

(c) A3 asphaltene

In the absence of magnetic field, the RDF curve shows multiple peaks at different distances where they represent multiple stackings (Figure 2-A3). The first clear and higher peak is at 4.5 Å, a second border peak is at 8.2 Å, and a third peak at 9.5 Å. Inspection of the molecular configurations showed that the first peak was due to face-to-face stackings and the second and third peaks represent two layers of parallel-displaced or T-shaped stacking molecules that were attached to the first peak. A3 asphaltene is an archipelago, its structure is more flexible than other used asphaltenes which could allow A3 to form multi-layer of stacking asphaltenes.

Applying magnetic field showed that 0.5 T could increase the aggregation affinity to form a large cluster of molecules, Figure 5. Unlike to A1 and A2, increasing magnetic field strength to 1.0 T and 1.5 T did not show any reduction in A3 aggregation. The response of A3 to the magnetic fields showed that not all asphaltene aggregations could be controlled by magnetic fields.

(d) A4 asphaltene

In the absence of magnetic field, the RDF curve shows multiple peaks at different distances where they represent multiple stackings (Figure 2-A4). The first clear and high two peaks are at 6.3 Å and 7.6 Å followed by a wider peak that exceeded 10 Å. Inspection of the molecular configurations showed that the first peak was due to parallel-displaced stackings and the second peak was because of a compensation of parallel-displaced and T-shaped stackings. Moreover, the other peaks were due to the closely attached dimers and trimers. Even with the long side chains, A4 asphaltene showed moderate aggregation affinity comparison with A3 with some structural hinderance due to flexible sulfide chains.

In a similar observation of previous asphaltenes, applying 0.5 T to the system increased the aggregation affinity and produced a large aggregate. Also, it changed the molecular configurations in comparison with 0.0 T and showed a high and border peak at 5.3 Å where the inspections showed it was due to ca ombination of face-to-face and parallel-displaced stackings. The same trend of A3 aggregation was observed in the case of A4. Increasing magnetic field strength than 0.5 T did not show any reduction in A4 aggregation, Figure 5. It increased the aggregation and changed the molecular configurations and stacking distances which may be attributed due to the effects of higher fields on the side chains and their flexibility.

3.2 Hydrogen bond

In this study, hydrogen bond interactions predominantly occur among the polar groups that have the ability for HB interactions between asphaltene molecules[23] (o-xylene and n-heptane do not have polar sites for such interactions). In asphaltene, the highly charged atoms are polar functional groups such as amino and carboxylic groups. The used asphaltenes have different sites that could participate in HB interactions; thus magnetic fields may have effects on the orientation and interactions of these atoms which may affect asphaltene aggregation.

The average number of HB between asphaltene molecules at different magnetic fields are reported in Figures 3. It shows the magnetic field effects on HB interactions where it caused an increase and decrease in the average number of HB interactions. The changes in HBs are due to the changes in the separation distances and/or the orientations of asphaltene molecules. Note that there is no clear trend in HB counts with respect to magnetic strength. This may strongly indicate the magnetic field strength predominantly affect the main aromatic systems of asphaltene and that the change in HB counts is merely due to the weak coupling of geometrical rearrangement of these aromatic systems with respect to the side chains where HB interactions occur.

3.3 Transport properties

The diffusivities of asphaltene have high importance in identifying molecular behavior. It is reported that variation in asphaltene diffusivities is due to their molecular size, structure, and the polarity. The heavier and high polar asphaltenes have lower diffusivities[14]. Further, magnetic field reduced the diffusivity of asphaltene as reported in Figure 4. In the absence of magnetic field, calculated diffusion coefficients are consistent with reported the asphaltene diffusivities. In the presence of magnetic fields, the reduction in asphaltene diffusivities was different depending on the asphaltene molecular structure and the strength of the applied field. The results suggested that asphaltene diffusivity could be controlled by applying an appropriate magnetic field. In addition, it is reported that asphaltene aggregation reduces asphaltene diffusion coefficient[14]. However, some of the results showed that the changes in the diffusivities were independent of aggregation status.

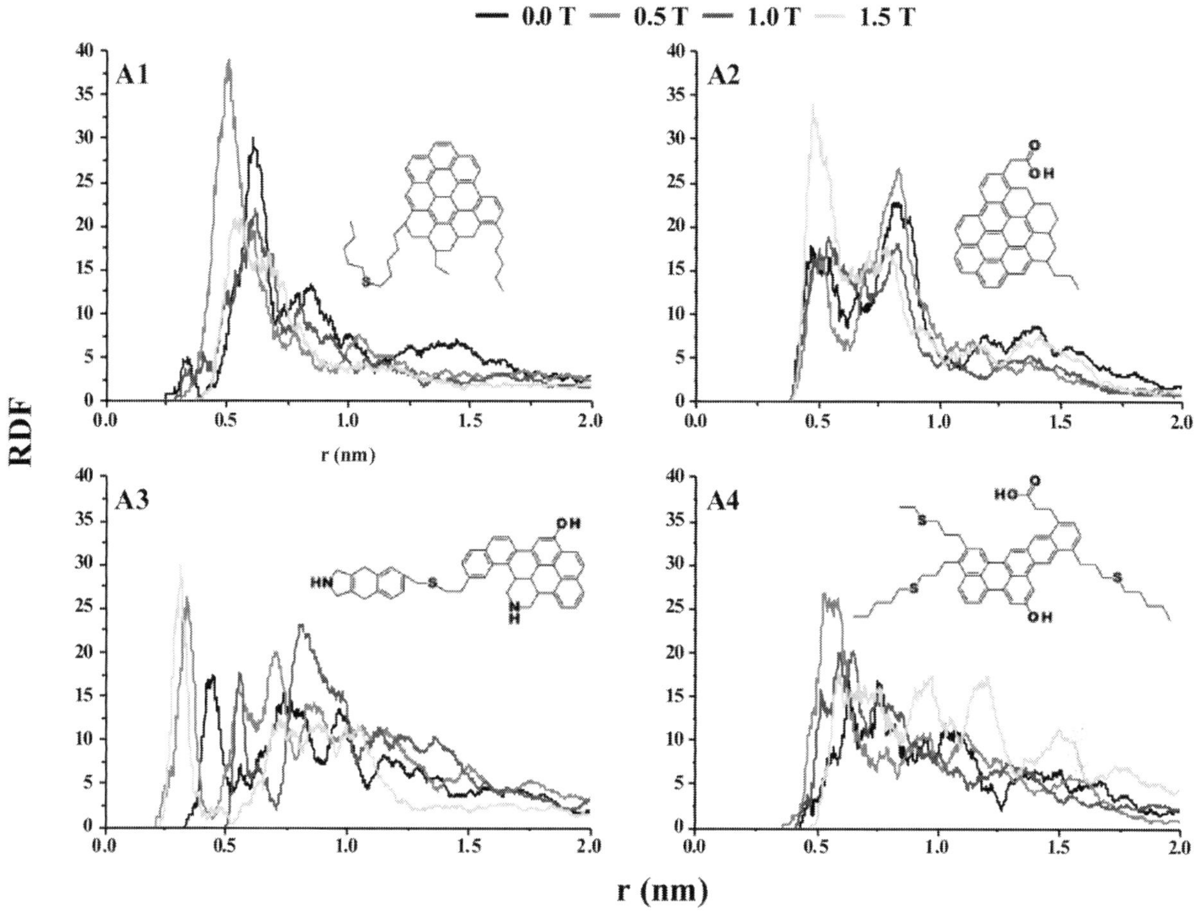

Figure 2. RDF curves of A1-A4 asphaltenes at different magnetic fields.

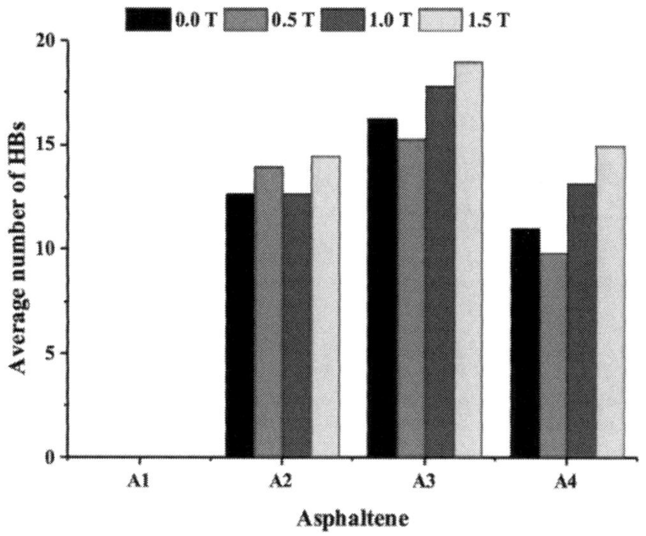

Figure 3. Average number of hydrogen bonds of A1-A4 asphaltenes at different magnetic fields (A1 has no sites for HB interactions).

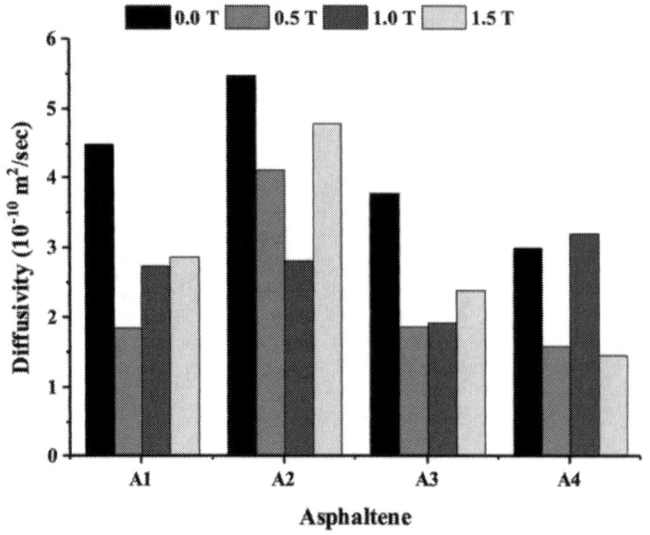

Figure 4. Diffusivities of A1-A4 at various magnetic fields.

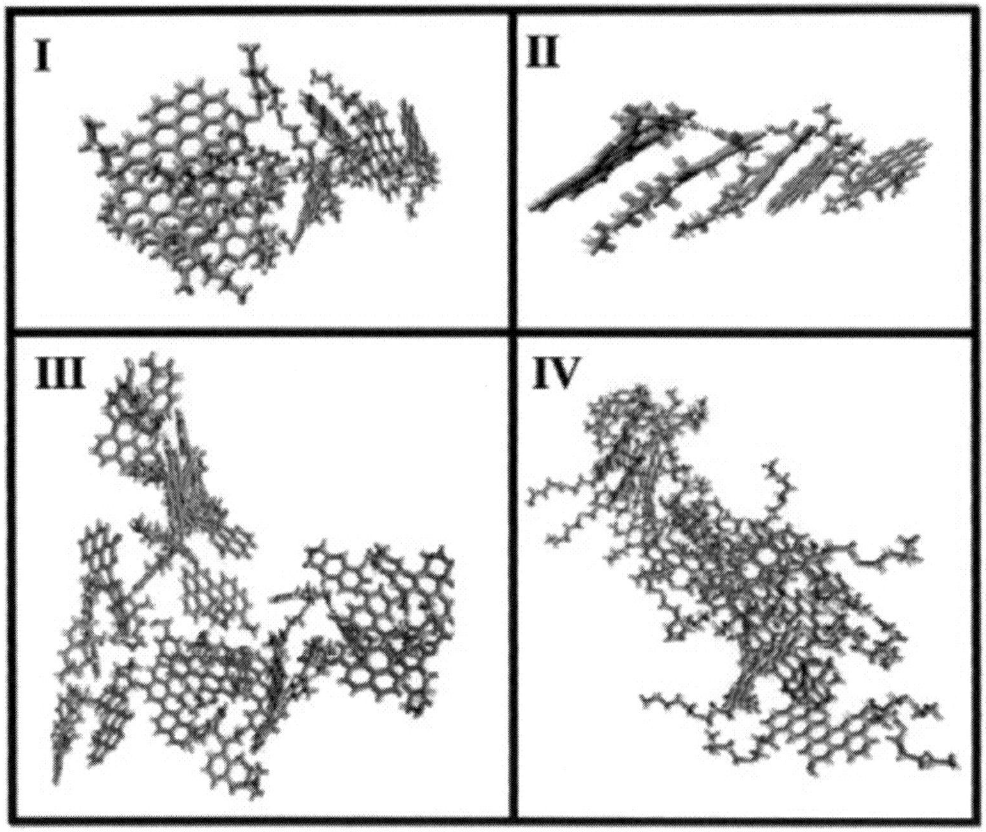

Figure 5. Snapshots of different asphaltenes at different magnetic fields. Where: I represents A1 at 1.0 T, II represents A2 at 0.0 T, III represents A3 at 0.5 T, and IV represents A4 at 1.5 T.

4. Conclusions

The effects of different magnetic fields concerning the asphaltene aggregation behavior in an equimolar mixture of heptane and xylene have been investigated. Radial distribution functions, hydrogen bond interactions, and the diffusivities were studied through molecular dynamics simulation. MD results showed different behavior of different asphaltenes. The different responses of asphaltenes could be explained as magnetic fields have a high effect on polar species and the outcomes become higher with increasing the polarity of the molecules. The results showed 0.5 T increased the aggregation and changed the preferred stackings. Besides that, the impact of 1.0 T, as well as the effect of 1.5 T, were different depending on the asphaltene molecular structure. Hydrogen bond results indicated that magnetic field did change the average number of the interactions depending on asphaltene architecture. These changes are due to the changes in the orientation and the distances between asphaltene molecules (as reported in RDFs). Moreover, the diffusivities of used asphaltenes were influenced by magnetic field. Magnetic field reduced the diffusivities of all the used asphaltene even when they had low aggregation status. It was observed that asphaltene with large single aromatic core changed more than others during magnetic treatments and the archipelago structure was less affected. This study clearly shows the complex relationship of the topological structures of asphaltene concerning magnetic field strength. The effects of magnetic field are totally dependent on the strength of the applied field, the medium, and the architecture of asphaltene.

Acknowledgment

This research is supported, in part, by the Higher Committee for Education Development in Iraq (HCED)/Office of the Prime Minister.

The authors acknowledge the Advanced Cyberinfrastructure for Education and Research (ACER) group at The University of Illinois at Chicago for providing HPC resources that have contributed to the research results reported within this paper. URL: https://acer.uic.edu

References

1. Luo, P. and Gu, Y. Effects of asphaltene content on the heavy oil viscosity at different temperatures. *Fuel* 86, 1069–1078 (2007).
2. Leontaritis, K.J. and Mansoori, G. A. A colloidal model for asphaltene flocculation from petroleum fluids. *Iranian Journal of Science & Technology* 16, 249–267 (1992).
3. Mousavi-Dehghani, S. A., Riazi, M. R., Vafaie-Sefti, M. and Mansoori, G. A. An analysis of methods for determination of onsets of asphaltene phase separations. *J. Pet. Sci. Eng.* 42, 145–156 (2004).
4. Kawanaka, S., Park, S. J. and Mansoori, G. A. Organic Deposition From Reservoir Fluids: A Thermodynamic Predictive Technique. *SPE Reserv. Eng.* 6, 185–192 (1991).
5. Mohammed, S. and Gadikota, G. The role of calcite and silica interfaces on the aggregation and transport of asphaltenes in confinement. *J. Mol. Liq.* 274, 792–800 (2019).
6. Gonçalves, J. L. *et al.* Study of the factors responsible for the rheology change of a Brazilian crude oil under magnetic fields. *Energy and Fuels* 25, 3537–3543 (2011).
7. Tao, R. and Xu, X. Reducing the viscosity of crude oil by pulsed electric or magnetic field. *Energy and Fuels* 20, 2046–2051 (2006).

8. Khalaf, M. H., Mansoori, G. A. and Yong, C. W. Magnetic treatment of petroleum and its relation with asphaltene aggregation (an atomistic investigation). *J. Pet. Sci. Eng.* 176, 926–933 (2019).
9. Tung, N. P. *et al.* Studying the mechanism of magnetic field influence on paraffin crude oil viscosity and wax deposition reductions. *Soc. Pet. Eng.* 1–7 (2001). doi:10.2118/68749-MS
10. Marques, L. C. C. *et al.* Study of Paraffin Crystallization Process Under The Influence of Magnetic Fields and Chemicals. in *Latin American and Caribbean Petroleum Engineering Conference* 1–6 (Society of Petroleum Engineers, 1997). doi:10.2118/38990-MS
11. Loskutova, Y. V., Yudina, N. V. and Pisareva, S. I. Effect of magnetic field on the paramagnetic, antioxidant, and viscosity characteristics of some crude oils. *Pet. Chem.* 48, 51–55 (2008).
12. Khalaf, M. H. and Mansoori, G. A. A new insight into asphaltenes aggregation onset at molecular level in crude oil (an MD simulation study). *J. Pet. Sci. Eng.* 162, 244–250 (2018).
13. Yaseen, S. and Mansoori, G. A. Molecular dynamics studies of interaction between asphaltenes and solvents. *J. Pet. Sci. Eng.* 156, 118–124 (2017).
14. Mohammed, S. and Gadikota, G. The influence of CO_2 on the structure of confined asphaltenes in calcite nanopores. *Fuel* 236, 769–777 (2019).
15. Mohammed, S. and Mansoori, G. A. Molecular insights on the interfacial and transport properties of supercritical CO_2/brine/crude oil ternary system. *J. Mol. Liq.* 263, 268–273 (2018).
16. Mohammed, S. and Mansoori, G. A. The Role of Supercritical/Dense CO_2 Gas in Altering Aqueous/Oil Interfacial Properties: A Molecular Dynamics Study. *Energy & Fuels* 32, 2095–2103 (2018).
17. Khalaf, M. H. and Mansoori, G. A. Asphaltenes aggregation during petroleum reservoir air and nitrogen flooding. *J. Pet. Sci. Eng.* 173, 1121–1129 (2019).
18. Yong, C. W. Descriptions and Implementations of DL-F Notation: A Natural Chemical Expression System of Atom Types for Molecular Simulations. *J. Chem. Inf. Model.* 56, 1405–1409 (2016).
19. Todorov, I. and Smith, W. The DL POLY 4 user manual. *Science & Technology Facilities Council (STFC).* (2016).
20. Della Valle, E. *et al.* Magnetic molecular dynamics simulations with Velocity Verlet algorithm. in *2017 32nd General Assembly and Scientific Symposium of the International Union of Radio Science, URSI GASS 2017* 2017–Janua, 1–4 (IEEE, 2017).
21. Yong, C. and Todorov, I. DL_ANALYSER Notation for Atomic Interactions (DANAI): A Natural Annotation System for Molecular Interactions, Using Ethanoic Acid Liquid as a Test Case. *Molecules* 23, 36 (2017).
22. Humphrey, W., Dalke, A. and Schulten, K. VMD: Visual molecular dynamics. *J. Mol. Graph.* 14, 33–38 (1996).
23. Yaseen, S. and Mansoori, G. A. Asphaltene aggregation due to waterflooding (A molecular dynamics study). *J. Pet. Sci. Eng.* 170, 177–183 (2018).

PHASE BEHAVIOUR OF HEAVY OIL AND *n*-BUTANE MIXTURES

Yulman A. Perez, Florian F. Schoeggl, Harvey W. Yarranton
University of Calgary, Calgary, Alberta, Canada

Introduction

Solvent-based methods are a potential alternative to thermal methods for the recovery of heavy oils because they are less water and energy intensive. Many proposed solvent-based recovery processes are designed to operate near the saturation pressure to maximize the solubility of the solvent in the oil while avoiding the formation of a second liquid phase. *n*-Butane is a solvent of interest for heavy oil recovery processes because it has a saturation pressure at typical operating temperatures that is close to the reservoir pressure of many heavy oil reservoirs. *n*-Butane is also a relatively low cost solvent. The injection of *n*-butane into the reservoir can lead to complex phase behaviour including the formation of liquid-liquid, vapour-liquid, and vapour-liquid-liquid regions. The second liquid phase may appear as a mobile liquid or a dispersed glassy asphaltene-rich phase depending on the conditions. The objectives of this study were to: 1) collect data for *n*-butane diluted bitumen at temperatures from 20 to 230°C and pressures up to 10 MPa; 2) tune a Cubic Equation of State (CEoS) with composition dependent mixing rules to match the data.

Experimental Methods

The experimental methodology is illustrated in Figure 1. Saturation pressures, liquid-liquid boundaries, phases compositions were measured for mixtures of *n*-butane and a Western Canadian bitumen provided by Shell. The vapour-liquid (VL_1) boundary was determined in a blind cell apparatus using the constant composition expansion method at temperatures from 50 to 230°C and *n*-butane contents from 3 to 50 wt%. The liquid-liquid (L_1L_2) boundary was determined using *n*-butane titration in a high pressure microscope (HPM) at temperatures from 50 to 180°C and pressures up to 10 MPa. The onset point was defined as the *n*-butane content at which the second liquid phase appeared. The second phase was a heavy asphaltene-rich phase and its morphology was examined visually in the HPM.

Following the terminology of deasphalting processes, the heavy phase is referred to as "pitch". Here, "pitch*" refers to the solvent-free pitch; that is, the bitumen in the pitch phase. The mass of each phase was determined directly in the PVT cell experiments or from a mass balance in the blind cell experiments. Phase compositions in terms of *n*-butane, maltenes, and *n*-pentane-insoluble asphaltenes (C5-asphaltenes) were measured in a DB Robinson Jefri PVT cell apparatus at a temperature of 130°C, a pressure of 10 MPa, and *n*-butane contents from 48 to 90 wt%. The C5-asphaltene and pitch* yields were measured in both a blind cell and a PVT cell apparatus at temperatures from 20 to 180°C and pressures up to 10 MPa. Yield is defined as the mass of the specified material in the pitch phase divided by the mass of bitumen in the feed. A more detailed explanation of experimental procedure can be found elsewhere (1, 2).

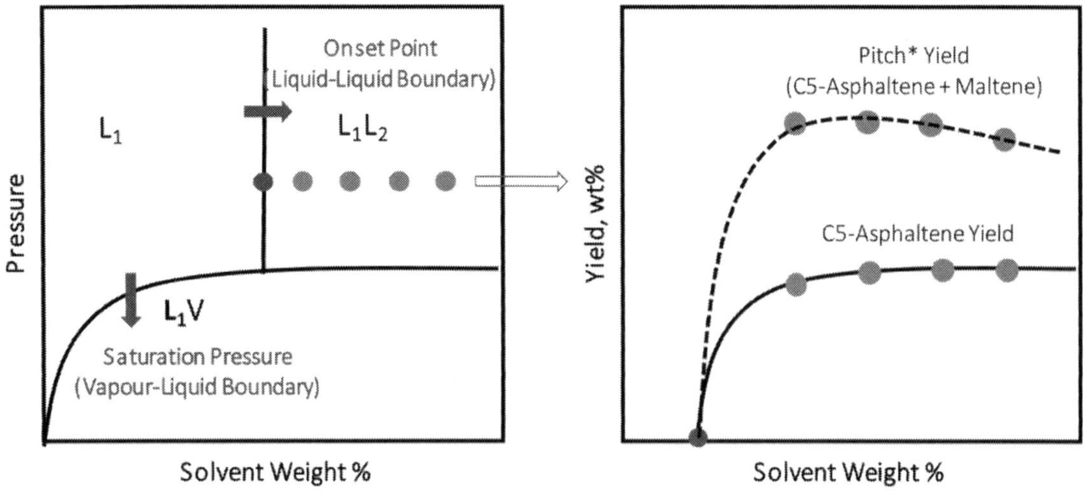

Figure 1. Phase diagrams showing the experiments performed on mixtures of *n*-butane and bitumen.

Heavy Oil Characterization and Modeling

A schematic of the heavy oil characterization methodology is provided in Figure 2 and details are reported elsewhere (1, 2). Briefly, the maltenes and C5-asphaltene fractions were characterized separately. The maltenes were divided into pseudo-components based on distillation data with a Gaussian extrapolation. The C5-asphaltenes were treated as a single pseudo-component. The critical properties and acentric factor for each pseudo-component were calculated from well-established correlations. These properties were input into the Advanced Peng-Robinson Equation of State (APR-EoS) implemented in VMGSim^{TM} (3).

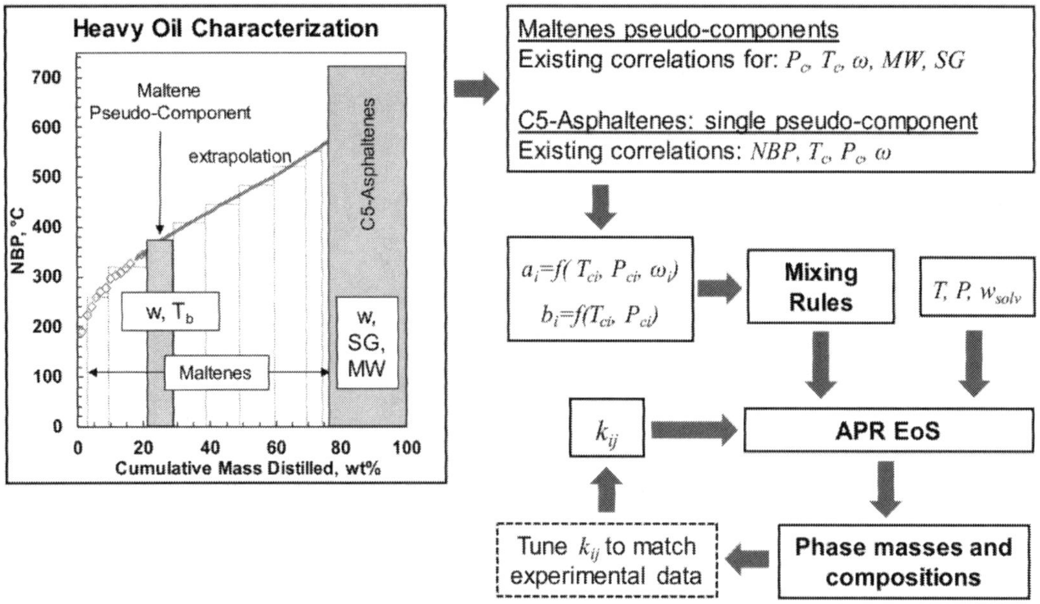

Figure 2. Schematic of characterization and modeling procedure for heavy oil/solvent phase behavior.

The Advanced Peng-Robinson equation of state (APR-EoS) as implemented in VMGSim™ (3) simulation software is given by:

$$P = \frac{RT}{v-b} - \frac{a\,\alpha(T_r,\omega)}{v(v+b)+b(v-b)} \tag{1}$$

where P is the absolute pressure, v is the molar volume, T is the absolute temperature, R is the universal gas constant, a and b are functions of the critical temperature and pressure and α is a function of the acentric factor. The APR-EoS includes volume translation.

The APR-EoS treats a mixture as a single component fluid with parameters a_{mix} and b_{mix} which in this study were determined from the classical van der Waals mixing rules given by:

$$a_{mix} = \sum_i \sum_j x_i x_j \sqrt{a_i a_j}(1-k_{ij}) \tag{2}$$

$$b_{mix} = \sum_i x_i b_i \tag{3}$$

where x_i and x_j are the mole fraction of components i and j, respectively, and k_{ij} is a binary interaction parameter. The binary interaction parameters were fitted to the experimental saturation pressure and yield data with the two alternative approaches as described below.

Temperature Dependent Binary Interaction Parameters (k_{ij})

Temperature dependent binary interaction parameters were proposed by Agrawal *et al.* (4) and are given by:

- maltenes with each other or with solvent (*n*-butane):

$$k_{ij} = k_{ij}^0\left(1 + \frac{k_{ij}^1}{T} + k_{ij}^2 l\,n(T)\right) \tag{4}$$

- C5-asphaltenes with maltenes or solvent:

$$k_{ij} = k_{ij}^0(1 + a_A T + b_A T^2) \tag{5}$$

where k_{ij}^0 is a reference interaction parameter proposed by Gao (5) and given by:

$$k_{ij}^0 = 1 - \left[\frac{2\sqrt{T_{ci}T_{cj}}}{T_{ci}+T_{cj}}\right]^n \tag{6}$$

where T_{ci}, T_{cj} are the critical temperatures of paired component i and j, respectively. The exponent n is an adjustable parameter set to the default value of 0.27, as recommended by Agrawal *et al.* (4). The parameters used to fit the data with the temperature dependent correlation are summarized in Table 1.

Composition Dependent Binary Interaction Parameters (k_{ij}*)

The temperature dependent binary interaction parameters were used as described above except for the binary interaction parameters between *n*-butane and the heaviest oil pseudo-components (Maltene8, Maltene 9, Maltene 10, and C5-Asphaltene). These pseudo-components were chosen to match the mass of the pitch* yield. The composition dependent binary interaction parameters for these components were calculated as follows (2):

$$k_{ij}^{*} = a + b(1 - e^{-6.5(w_{solv} - w_{LL})}) \tag{7}$$

$$a = a_1 NBP + a_2 \tag{8}$$

$$b = b_1 NBP + b_2 \tag{9}$$

with the following constraints:

$$\text{If } k_{ij}^{*} > 0, k_{ij} = k_{ij}^{*}$$

$$\text{If } k_{ij}^{*} \leq 0, k_{ij} = 0$$

where NBP is the pseudo-component normal boiling point in °C and w_{solv} is the solvent (*n*-butane) mass fraction in the feed. The parameter w_{LL} is set equal to the measured mass fraction of *n*-butane at the onset of asphaltene precipitation. The parameters a_1, a_2, b_1, and b_2 are constants and were tuned to fit the measured C5-asphaltene and pitch* yield data using a least squares regression. The fitted parameters are provided in Table 2.

Table 1. Tuned parameters for the temperature dependent parameter correlations (temperature in K).

Pseudo-Component Pair	k_{ij}^{1} K	k_{ij}^{2}	a_A K^{-1}	b_A K^{-2}
Maltenes	443.50	0.00809	-	-
C5-Asphaltene	-	-	-0.00747	0.0000262

Table 2. Tuned parameters for the composition dependent parameter correlations.

Parameter	Value
w_{LL}	0.41
a_1	0.000077029 °C^{-1}
a_2	0.00001011
b_1	0.00010358 °C^{-1}
b_2	0

Results

Measured Phase Behaviour

The measured data were used to generate pressure-composition (P-X) phase and yield diagrams. Figure 3a shows a typical P-X diagram in this case at 130°C. At low *n*-butane contents, the mixtures of *n*-butane and bitumen formed a vapour-liquid (VL_1) phase at low pressure and a single liquid phase (L_1)

above the saturation pressure. The saturation pressure increased monotonically with increasing *n*-butane content until approximately 40 wt% *n*-butane. Above this *n*-butane content, the mixtures formed two liquid phases (L_1L_2) at pressures above the saturation pressure and a vapour-liquid (VL_1) region below the saturation pressure. The L_1/L_1L_2 boundary was insensitive to temperature and only slightly sensitive to pressure (a vertical line on the P-X diagram). The $L_1L_2/V\ L_1L_2$ boundary was approximately equal to the vapour pressure of *n*-butane (a horizontal line on the P-X diagram). In the L_1L_2 region, the C5-asphaltenes all precipitated immediately above the onset, Figure 3b. The pitch* yield reached approximately 40 wt% just above the onset and decreased slightly at higher *n*-butane contents, Figure 3b and other yield data not shown here. The yield data demonstrate that the L_1L_2 region extends to a feed *n*-butane content of at least 90 wt%.

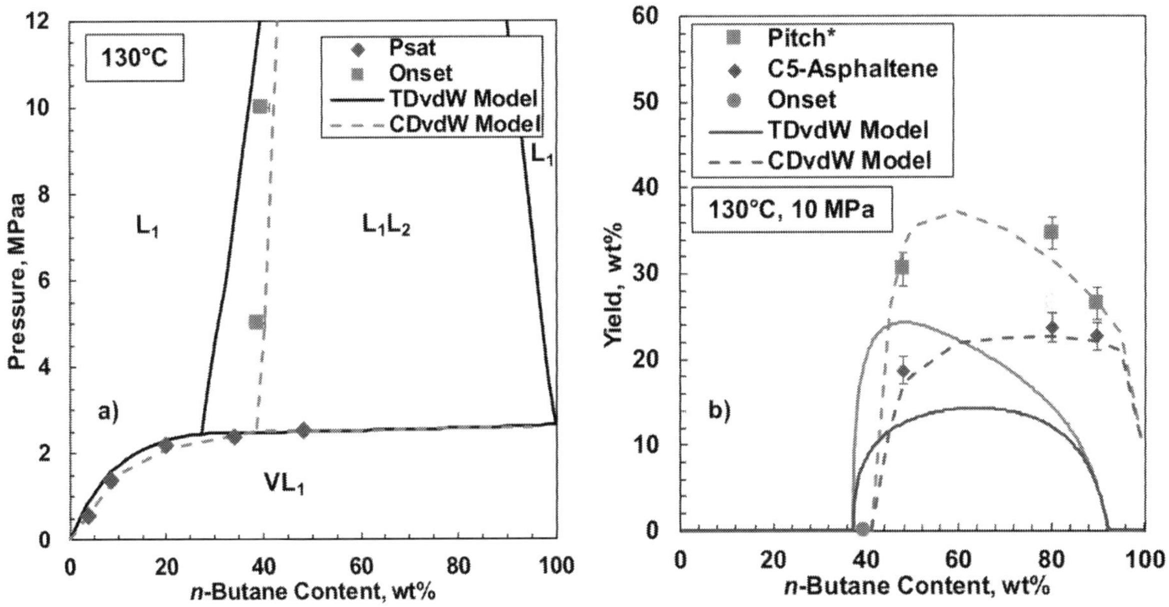

Figure 3. Measured (PVT cell) and modeled phase behaviour of *n*-butane and a Western Canadian bitumen: a) pressure-composition diagram at 130°C; b) yields at 130°C and 10 MPa.

The pitch phase onsets for *n*-butane diluted WC-B-B5 bitumen are provided in Table 3. The average onset is 39.6 wt% *n*-butane content and is insensitive to temperature and pressure. This value is consistent with mixtures of bitumen with different carbon number *n*-alkanes. For example, the onsets for bitumen mixed with propane and *n*-pentane are approximately 30 wt% (2) and 50 wt% (1), respectively.

The pitch phase morphology was analyzed in a high pressure microscope (HPM). The pitch phase was opaque and the solvent-rich phase was translucent at conditions evaluated. At all temperatures and pressures considered in this study, the pitch phase appeared as small particles or droplets just above the onset as shown in Figures 4a and 4c. At 50 and 90°C, the pitch phase became more liquid-like or possibly was in a glass transition state with increasing *n*-butane content, and the particles/droplets aggregated, Figure 4b. At 130°C, the pitch phase was a liquid that coalesced and settled into a distinct layer as the *n*-butane content increased, Figure 4d.

Table 3. Onset of the heavy pitch phase in mixtures of *n*-butane and bitumen. The average uncertainty in the onset ± 1.6 wt% *n*-butane.

Temperature °C	Pressure MPa	*n*-Butane Content wt%
50	5	39.2
0	5	41.0
130	5	38.7
50	10	38.5
90	10	41.0
130	10	39.6
180	10	40.5

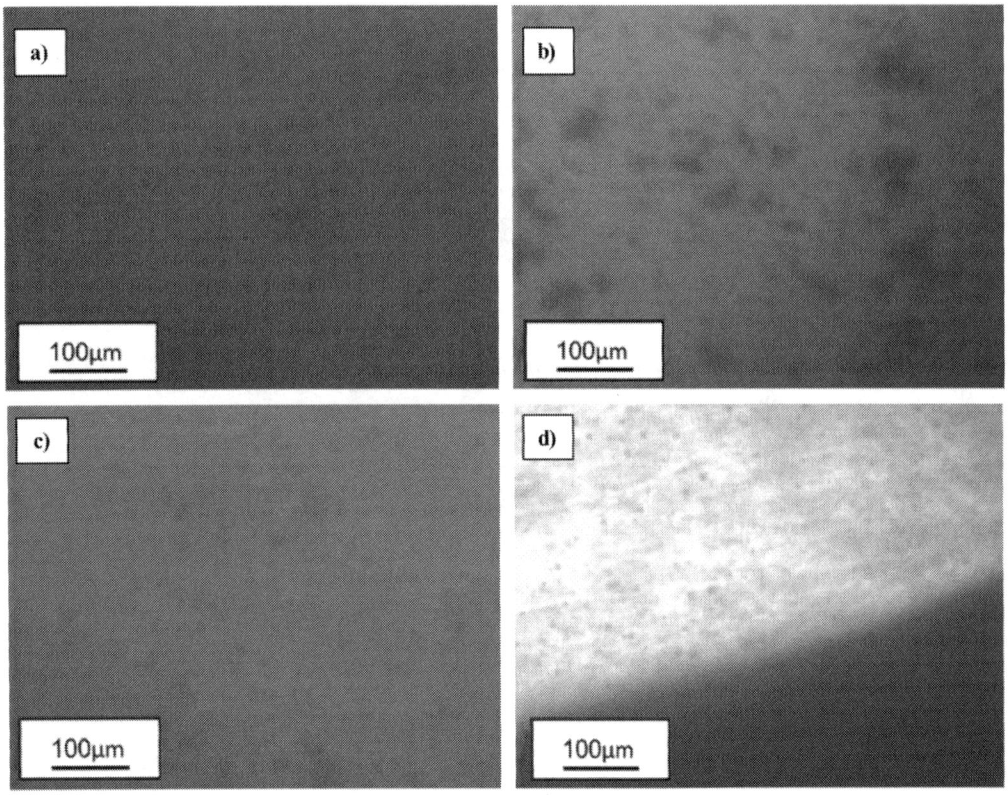

Figure 4. HPM micrographs of the L_1L_2 region in mixtures of *n*-butane and a Western Canadian bitumen at: a) 50°C, 5 MPa, and 41.9 wt% *n*-butane; b) 50°C, 10 MPa, and 42.3 wt% *n*-butane; c) 130°C, 5 MPa, and 40.2 wt% *n*-butane; d) 130°C, 10 MPa, 52.8 wt% *n*-butane. The pitch phase is dark and the *n*-butane-rich phase is light.

Phase compositions were analyzed in terms of the *n*-butane, maltene, and C5-asphaltene contents for both the solvent-rich and pitch phase in the L_1L_2 region. Figure 5 shows the phase compositions from the PVT cell at 130°C, 10 MPa, and different *n*-butane contents. Almost all of the asphaltenes were rejected from the solvent-rich phase even near the onset. The *n*-butane content in the pitch phase was approximately 20 wt% near the onset and decreased monotonically towards near zero at high dilution. In general, approximately half of the bitumen partitioned to the pitch phase. This behaviour is consistent with mixtures of bitumen with different carbon number *n*-alkanes. For example, for bitumen mixed with propane, more than half of the bitumen partitioned to the heavy (pitch) phase (2). For bitumen mixed with *n*-pentane, the heavy phase consisted mainly of asphaltenes and the majority of the bitumen remained in the solvent-rich light phase (1).

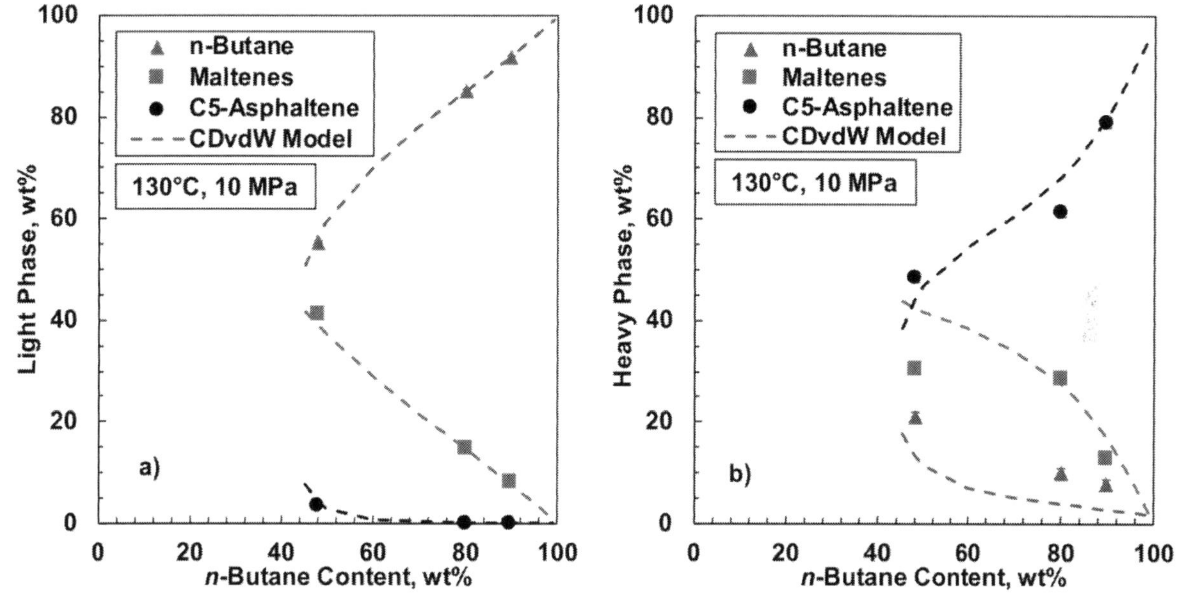

Figure 5. Phase composition of *n*-butane diluted WC-B-B5 bitumen at 130°C and 10 MPa for: a) light (solvent-rich) phase; b) heavy (pitch) phase.

Modeled Phase Behaviour

The APR EoS with the temperature dependent interaction parameters (TDvdW) generally fit the saturation pressures and onsets within the experimental error, Figure 3a. However, larger deviations were found at temperatures higher than the critical temperature of the *n*-butane (152°C). The TDvdW approach predicted a non-existent L_1L_2/L_1boundary above 90 wt% *n*-butane content and significantly under-predicted the C5-asphaltene and maltene contents in the pitch phase. The APR EoS with composition dependent parameters (CDvdW) fit not only saturation pressure and onsets but also the phase masses generally to within the experimental error, Figure 3b. The CDvdW approach fit the experimental data for the solvent-rich phase compositions to within the experimental error of ±0.3 wt%, Figure 5a. This approach predicted the correct trends for the pitch phase compositions but underpredicted the *n*-butane content in the pitch phase by approximately 6.44 wt%, Figure 5b.

Conclusions

The phase behaviour of heavy oil and *n*-butane mixtures were examined at temperatures from 20 to 230°C and pressures up to 10 MPa. Both vapour-liquid (VL_1) and liquid-liquid ($L_1 L_2$) equilibria were observed. The onset (L_1/L_1L_2) boundary appeared at 40 ±1 wt% *n*-butane content in the feed and was insensitive to temperature and pressure. In general, approximately half of the bitumen partitioned to the heavy "pitch" phase. The pitch phase appeared as glassy particles near the onset at 50 and 90°C but became a liquid by 130°C. The APR EoS with composition dependent van de Waals mixing rules (CDvdW) fit the measured phase boundaries, yields, and solvent-rich phase compositions to within experimental error and the pitch phase compositions to within ±5.3 wt%.

Acknowledgments

The authors are grateful for financial support from the sponsors of the Natural Science and Engineering Research Council of Canada (NSERC) Industrial Research Chair in Heavy Oil Properties and Processing, including NSERC, CNRL, CNOOC Nexen, Ecopetrol, Schlumberger, Suncor Energy, and Virtual Materials Group.

References

1. Johnston, K. A., Schoeggl, F. F., Satyro, M. A., Taylor, S. D., & Yarranton, H. W. (2017). Phase behavior of bitumen and n-pentane. Fluid Phase Equilibria, 442, 1-19.
2. Mancilla-Polanco, A., Johnston, K., Richardson, W. D. L., Schoeggl, F. F., Zhang, Y. G., Yarranton, H. W., & Taylor, S. D. (2018). Phase Behavior of Heavy-Oil/Propane Mixtures. Society of Petroleum Engineers.
3. Virtual Materials Group Inc VMG. (2011). VMG Sim User's Manual. VMGSim, Version 9.5, Calgary, Canada.
4. Agrawal, P., Schoeggl, F. F., Satyro, M. A., Taylor, S. D., & Yarranton, H. W. (2012). Measurement and modeling of the phase behavior of solvent diluted bitumens. Fluid Phase Equilibria, 334, 51-64.
5. Gao, G., Daridon, J. L., Saint-Guirons, H., Xans, P., and Montel, F. (1992). A simple correlation to evaluate binary interaction parameters of the Peng-Robinson equation of state: binary light hydrocarbon systems. Fluid phase equilibria, 74, 85-93.6.

550317 Effect of Reheating on the Flowability of Waxy Crude Treated with Pour Point Depressant

Waxy crude oils usually present high pour points and poor flowability at ambient temperature, causing significant flow assurance challenges in oil production and transportation. Addition of pour point depressant (PPD) is known as an effective approach to improve the cold flowability of waxy crudes. It is recently discovered that the efficacy of PPD as determined by conventional pour point tests depends on the thermal history of the treated crude oil during the test.

This paper is dedicated to investigating the effect of sample temperature manipulation, especially the reheating operation after PPD treatment, on the measured pour point of treated crude. To achieve this goal, three crude oils A, B and C, treated with 100 ppm of a PPD at temperatures sufficiently higher than their corresponding wax appearance temperatures, were equilibrated at room temperature, reheated to a temperature in the range from 30 to 70 °C, and eventually cooled for pour point determination.

We discovered that reheating the treated sample to a temperature (T_{reheat}) between the wax appearance temperature (WAT) and the wax dissolution temperature (WDT), i.e., WAT $< T_{reheat} <$ WDT, causes the efficacy of the PPD to diminish, while reheating to a temperature outside this range, i.e., $T_{reheat} <$ WAT or $T_{reheat} >$ WDT does not significantly deteriorate the efficacy of PPD.

When the three treated oils are subjected to reheating to a temperature in the undesired range between WAT and WDT, the measured pour points of the treated oils are 7 °C, 8 °C and 7 °C higher than the corresponding lowest pour points achievable with the same treatment. In the worst-case scenario, reheating can completely nullify PPD treatments. The deterioration of PPD efficacy due to undesired reheating operation is also corroborated by increases in the viscosity and yield strength. In addition, after undesired reheating, the wax particles precipitated upon cooling form an entangled matrix featuring low fractal dimensions and small crystal sizes. This wax microstructure differs from those observed after treatment but in absence of improper reheating while resembles the microstructure of wax particle network generated with an untreated oil.

Based on this investigation, we recommend that caution be exercised when assessing the efficacy of PPD's with conventional pour point test methods. Special care is required when the reheating temperature of 45 °C defined by the ASTM D5853 method falls in the range between the WAT and WDT of the oil of interest, as reheating to 45 °C will result in diminished efficacy of PPD under this circumstance.

This paper is supported by the National Natural Science Foundation of China (No.51534007, No.51134006 and No.51304224).

551073 A Conceptual QCM-Study of Formation, Aggregation, Adsorption and Deposition of Asphaltene Particles

Formation and deposition of asphaltene, considered as the heaviest and the most polarizable fraction of crude oil, can cause severe problems in oil production and extraction systems. Production companies bear huge reduction or even cessation in production and, in turn, un-avoided costs due to asphaltene precipitation, aggregation and deposition in wellbore or production tubings. To gain more and better insight into asphaltene formation in production systems, the conceptual investigation on the mechanisms of asphaltene adsorption and deposition on a solid surface is indispensable.

In this research work, the Quartz Crystal Microbalance (QCM) experimental technique was utilized to probe the kinetics of asphaltene adsorption and desorption using a gold surface electrode. To study the effects of flow conditions on the asphaltene aggregation and deposition, a novel and visual QCM dynamic flow cell was designed and constructed. It should be stated that throughout the experiments various model oil samples at different asphaltene contents were synthesized and the formation, aggregation and deposition of asphaltene particles were studied under different dynamic conditions using a high resolution camera, mounted on the top of the flow cell. The time-dependent size distribution of asphaltene aggregates inside the flow cell were also monitored throughout the experiments. The model oil samples were comprised of toluene, normal heptane and asphaltene particles, extracted from crude oils, at different concentrations ranging from 100 ppm to 1500ppm. The deposition of asphaltene particles on the gold surface was also investigated using the model oil containing n-heptane and toluene and the deposition rate constants reported for different samples at various static and dynamic conditions.

It was concluded that both asphaltene adsorption and deposition can increase with the concentration of asphaltene particles in the model oil samples. At higher injection rates, longer time was needed for the amount of asphaltene adsorbed on the surface to reach the steady state condition while this time was much shorter in the case of asphaltene deposition. It was also observed that while the asphaltene deposition can rapidly increase with the asphaltene concentration at small range of concentrations, it can decrease at higher asphaltene concentrations, mainly, due to the aggregation of asphaltene particles. In other words, at higher asphaltene concentrations the asphaltene particles tend to get aggregated and less asphaltene deposition can occur in the aggregation-dominated range of concentration.

551425 Real-Time Determination of Well and Formation Parameters Using Ensemble Kalman Filters

This paper discusses applications of the Ensemble Kalman Filter (EnKF) for the real-time quantification of sensor-generated data in two examples: the analysis of the declining production curve and zonal pressure sensor data for evaluating the matrix permeabilities and processing of multichannel optical to monitor the cleanup of hydrocarbon fluid samples in oil-based muds during formation-tester sampling.

The EnKF algorithm is an elegant and effective method used to optimize model parameters based on differences in predictions of model and measurement data. This paper explains how the same EnKF algorithm can be successfully applied to segmented multichannel sensor data in two very different applications: declining total production and zonal pressure data in a synthetic example of a fracture-stimulated wellbore, and field data obtained from multichannel optical density sensors showing the gradual transition from oil-based mud (OBM) filtrate to native formation fluids during formation-tester sampling stations. In the first case, the pressure depletion and production is simulated using an effective 2-D fracture production model, while a simple algebraic proxy model is used to predict the decline of the volumetric fraction of OBM filtrate with time during formation-tester sampling.

To test the algorithm, a MATLAB code was developed, proving the high efficiency of the suggested method. At each time step, approximately 100–150 sampling data points from the multichannel sensors were processed. Synthetic (simulated) pressure flow rate data was used in the production decline case, while the actual field data from 8 channel optical sensors were used in the formation-testing case. Model runs were performed in 50–60 combinations of the model parameters, which were normally distributed around the best-guess values at the initial step. In both cases, only 2–3 iterations of the algorithm were sufficient to obtain the values of the matching parameters.

In the case of the fracture system, the zonal permeabilities, which varied within an order of magnitude, were effectively found, with accuracy defined by the accuracy set for the measured pressure and flow rates. In the formation-testing case, the spectral optical densities, instant values of the OBM volume fraction, and the endpoints (spectral densities of mud and oil) were determined. The algorithm has shown to converge for all test runs, and an average accuracy of approximately 5 percent has been achieved.

Currently, both dynamic history matching for fracture modeling and formation-tester analysis are being carried out manually, making real-time analysis difficult. The suggested method combines advantages of modern multichannel data acquisition and advanced data science. The algorithm is universal, fast, stable, and can be applied to a variety of different workflows and models incorporating multichannel data and run in real time, thereby enhancing the operators' ability to make timely decisions during operations.

550333 An Integrated Urban PLAN for CO-Developing OIL & Gas Fields

This paper records the accomplished works, efforts, and collaborations between stakeholders whilst developing the integrated surface urban plan for three overlapped fields to secure and sustain the production targets. It also highlights the means of innovation, creativity, and thinking out of box to achieve a mutually agreed urban plan. Moreover, it outlines the success partnership story of Kuwait Oil Company and Shell Kuwait E&P using the Extended Technical Service Agreement Framework.

The Urban Plan is the conceptual capturing and definition of the surface facilities, gathering and exporting networks, as well as the supporting infrastructure such as electrical power, fencing, access for drilling rigs, etc. In the context of an oil and gas fields development, the urban plan focuses explicitly on the surface aspects of the life cycle development of the field. As a process, it involves strategic planning, technical evaluation, consultation, concept engineering, development planning, and collaboration between a host of stakeholders. Normally, it minimizes production deferment, optimizes cost for future developments, and proactively addresses HSSE risks and their mitigations.

Its associated challenges start with the assumed maturing mechanism(s) for a field(s) i.e. its development phases; production targets/plateaus; wells drilling sequence and patterns; in addition to reservoir uncertainties e.g. souring, and production, etc. The spectrum of the surface challenges embraces flow assurance; existing infrastructure; supply of utilities (water, power, fuel gas, etc.); current flowlines and pipelines networks; running facilities limitations; Logistics; and many more.

This urban plan is unique as it integrates the surface facilities and infrastructure of three distinguished gas, conventional oil, and heavy oil fields that are all sour and largely co-exist on a congested surface footprint though however, at different subsurface depths with reservoir boundaries. The uniqueness originates from the simultaneous development of the three fields with diverse recovery mechanisms (natural depletion, heavy oil cold production, EOR/IOR flood, etc.), and the co-existence of rigs to drill single and multiple wells pads, as well as the scattered early production facilities, arbitrary comingled pipelines networks, and temporary infrastructure escalate the potential complexity of developing the integrated urban plan and contribute to Simultaneous Operations (SIMOPS) concerns and challenges.

The key success factors were early engagement with stakeholders, harmonization between developing teams, cognitive flexibility, and synergy enabled the timely identification of common grounds and rational utilization of the surface footprint particularly with the massive well counts needed to drain the heavy oil field and the change of recovery mechanism in later developments.

The paper concludes the successful partnership and integration of multi- disciplines and multi-teams within Kuwait Oil Company and Shell Kuwait E&P resulting in an integrated master urban plan that hosts the short, medium, and long-term surface facilities, pipelines networks, and associated infrastructure that grant the co-production of the three fields whilst satisfying the vast maturation mechanisms; limited surface footprint; timely simultaneous developments; and HSSE guidelines.

550700 Understanding the Role of Iron Tosylate on Heavy Oil Viscosity Reduction

Heavy oil is an abundant energy resource, but its recovery remains challenging primarily due to its high viscosity. Thermally enhanced oil recovery in the presence of metal-ligand compounds (MLCs) has been studied as a promising method for *in situ* viscosity reduction and oil quality upgrading. In spite of its importance, the interactions between MLCs and crude oil components at the molecular-level are poorly understood, as well as, the mechanistic details for viscosity reduction are unknown. Here, we studied viscosity changes of real crude oil in the range of 80 - 295 °C with or without iron para-toluenesulfonate (tosylate) MLC, and analyzed reaction products via viscometry, TGA-MS, NIR spectroscopy, GC-MS, SARA analysis, XRF, and XPS. Thermal treatment lowered viscosity at all temperatures, whereas thermal treatment in the presence of this MLC decreased viscosity at temperatures above 230 °C. The MLC effect was strongest at 280 °C, at which thermal treatment without and with the iron tosylate decreased viscosity by 39% and 58%, respectively. 4-methylbenzenethiol was detected in the oil, corresponding to the loss of one of the three para-toluenesulfonate ligands. The MLC also lowered asphaltene fractions more than thermal-only. We propose the MLC lowers oil viscosity by releasing a ligand to form a compound that interferes with asphaltene interactions, and by catalytically reacting with the asphaltene which decreased the asphaltene fraction. This understanding of MLC-induced deviscosification provides better rationale for the selection of metal and ligand to improve thermally enhanced oil recovery.

AUTHOR INDEX

AIChE
120 Wall Street, FL 23
New York, NY 10005-4020

ISBN 978-1-5108-8678-0

9 781510 886780